Biomedical Response of Photoactive SiO2-Zn@Fe2O3 Nanofibers

Biomedical Response of Photoactive SiO2-Zn@Fe2O3 Nanofibers

Kamran Tahir
Zia Ul Haq
Sadia Nazir

ELIVA PRESS

Published by Eliva Press
Email: info@elivapress.com
Website: www.elivapress.com

ISBN: 978-1-63648-306-1

© Eliva Press, 2021
© Kamran Tahir, Zia Ul Haq, Sadia Nazir
Cover Design: Eliva Press
Cover Image: Freepik Premium
Printed at: see last page

Biomedical response of photoactive SiO$_2$-Zn@Fe$_2$O$_3$ nanofibers

Kamran Tahir

Abstract

Here, in the presence of Fe$_3$O$_4$ nano-fibers, we prepared SiO$_2$-Zn@Fe$_2$O$_3$ hybrid Nano-fibers through a novel and simple one-pot redox reaction between ZnSO$_4$ & SiO$_2$. The Fe$_3$O$_4$ exterior nano-fibers would be homogenously covered by SiO$_2$ coating to arrange a distinctive core-shell construction and then Zn nanoparticles are intercalated in the covering of SiO$_2$. The synthesized nanofibers were tested for photodegradation of methylene blue (MB). The result showed that 99% MB was degraded in 60 minutes. Furthermore, the antibacterial potential of SiO$_2$-Zn@Fe$_2$O$_3$ nanofibers was tested against *E. coli* and *S. aureus* bacteria both in light and dark. The impact of different analysis such as Reactive oxygen species (ROS) analysis, irradiation effect on bacterial inhibition, concentration effect of SiO$_2$-Zn@Fe$_2$O$_3$ nanofibers and reduction of DPPH studied. The findings clearly demonstrate that ROS is produced in the presence of SiO$_2$-Zn@Fe$_2$O$_3$ nanofibers in bacterial cells and is responsible for their inhibition. Findings have shown that syn-thesized nanostructures can also increase the stability of DPPH radicals with in-creasing concentrations of nanomaterials, making them a strong candidate for DPPH reduction. The overall results show that the efficacy of SiO$_2$-Zn@Fe$_2$O$_3$ nanofibers for inhibition was more pronounced than that of individual iron oxides.

Key words: Hydrothermal method; SiO$_2$-Zn@Fe$_2$O$_3$; Photodegradation of MB; Photoinhibition of bacteria

Contents

1. Introduction

Since their simplistic synthesizing mechanism, effective catalysis and distinguishing electronic state, zinc nanoparticles (Zn-NPs) has gained great consideration in the last three decades. These experiments have been used widely in a wide range of catalytic reactions, including alcohol and CO oxidation, nitroarene reduction, C-C coupling and dye reduction [1-5]. While immaculate Zn nanoparticles are shown to improve catalytic activity, it was difficult to distinguish from the reaction to recover nanoparticles, limiting its application of elegant chemical substances for industry construction. In addition, individual Zn- nanoparticles appear to aggregate at minute size in this catalysis process, greatly decreasing their activity. As a result, many attempts have been made to produce catalysts by the regulation of Zn-nanoparticles at different maintenances [6-10]. In specific, Zn-nanoparticles are preserved on the surfaces of some standard metal oxide. TiO_2, SiO_2 and ZnO typically provide a remarkably high degradation ratio when producing negatively charged Zn forms used to pass electrons from the oxides that cause the smaller particles, e.g. O_2 [11-13]. The continuous bonding of Zn-oxide is strongly dependent on the discrimination of methyl alcohol oxidation [14].

In the fields of cosmetics, garments, plastics, paint, and pigment, a considerable amount of numerous forms of dyes are used by advancing science & social progress, resulting in the creation of essential industrial exploitation of H_2O contamination [15]. A significant percentage of dyes are not biodegradable, where the ecological integrity of sea organisms, which is harmful to humans, should be interrupted [16-18]. The oxidation and removal of colorants is therefore a critical issue. Semiconductors like ZnO and TiO_2 etc are widely used as photocatalysts [19-20].

Various pathways are followed to recover these water soluble toxins i.e. dry-clean metallic & metal-organic frames [21-22].

Various chemical processes are studied in which noble metals are used as catalyst e.g. silver, platinum, etc [23]. But these metals are expansive. Thus Zn is the alternative of all the expansive metals due to its excellent catalytic activity and dye-reduction contributions. The Zn-nanoparticles are now-a-day sponsored by a lot of studies for the catalytic reduction of coloring. For example, Zn-nanoparticles retained the photocatalytic action for the deprivation of Rhodamine B (RhB) on silicon nano-wires [24]. Deng & co-workers synthesized MnO_2/Zn-nanoparticles, showing the better productivity decrease for MB compared to the separate materials. However, the preparation of successful Zn-NPs built for the collection of dye degradation is still an enormous challenge. Liu & coworkers only produce formaldehyde samples of Zn/Fe_3O_4@aminophenol, showing the strong catalytic productivity of positive MB dye and negative methyl orange (MO) dye [25]. Silica typically uses rare earth oxide, which is recognized as having distinctive active sites, known for its ability to differ in the oxidation state between +4 & +3, where it is considered the best supporting medium for Zn nanoparticles e.g, ZnSiO_2 Janus like nanoparticles will be invented & revealed an advanced operation in the direction of degeneration of MO as opposed to SiO_2 nanoparticles and other forms of Zn/SiO_2 [26, 27]. Zn/ZnO-SiO_2 nanomaterials are identified and used for MB reduction [28]. This would be an excellent choice for the assimilation of Zn-nanoparticles within the SiO_2 medium to create a nanostructure for the prevention of aggregation & leaching of Zn-nanoparticles. For powerful interactive among SiO_2 & Zn ingredients, this one is predictable to lift the catalytic competence [29]. Several years ago, Zn@SiO_2 nanostructure core shells were developed using various methods that were admirably extended to different forms of catalytic reactions [30-32].

While the manufacturing process of Zn-NPs and SiO_2, efficient catalytic activity and great selectivity, excellent recycling capability after dye degeneration is still a major challenge.

There are several categories of microbes on earth which affect living organism to cause infections or outcome in discoloration of textile and wastewater and create disgusting odors and tastes. The microorganisms comprise viruses, bacteria, parasites and algae. The shallow wells are the example of highly threat of impurity by microbes; specifically, those situated neighboring to the flora and fauna, farms & great danger flood areas. H_2O overflows in this area generally originate water pollution. Most of the bacteria show great resistance to the available antibiotic. Zn and ZnO are widely used as antibacterial agent with combination of other nanomaterials. ZnO has the ability to produce reactive oxygen species which play an important role in the inhibition of bacteria.

In the project, we have synthesized SiO_2-Zn@Fe_2O_3 nanofibers, is effectively prepared via a new hydrothermal process. The synthesized nanostructure was applied as a photocatalyst against methylene blue degradation and photo-deprivation against bacteria.

2. Materials and method

2.1 Synthesis of SiO_2-Zn@Fe_2O_3

The new modified hydrothermal procedure was used for the preparation of SiO_2-Zn@Fe_2O_3 Nanofibers. According to the procedure, 0.7 g of Fe $(NO_3)_3 \cdot 9H_2O$ was dissolved in a mixture of 15 ml of isobutyl alcohol and 15 ml of N,N-dimethylacetamide and stirred for 30 minutes. After that 2g of polyvinyl pyrrolidone was added, this will be stirred well at room temperature for 10 hr to form a transparent and homogenous mixture. Then added 10 ml of 0.05 M $CuCl_2$ and stirred for 30 minutes. In order to get the pH 10, 0.2 M (15 mL) NaOH was added to the above solution. Then 6 mL tetraethylorthosilicate (TEOS) was drop wise added to the above solution and again stirred for 2 hours. Then 5 mL tween-80 was dissolved in 30 mL water and added to the above solution, followed by stirring at 25 °C for 8 hours. After that a pH of 10 was maintained by the addition of 10 mL of 0.2 M NaOH and again stirred for 2 hours. The synthesized nanofibers were transferred to Teflon autoclave, heated at 120 °C for 2 hours. After hydrothermal treatment the synthesized nanomaterials were washed and dried in vacuum oven and calcined in air at 200 °C for 2 hours. The reddish brown powder confirmed the SiO_2-Zn@Fe_2O_3 nanofibers. The Fe_2O_3 nanofibers were synthesized by the same method without the use of Si, Zn salts and CTAB.

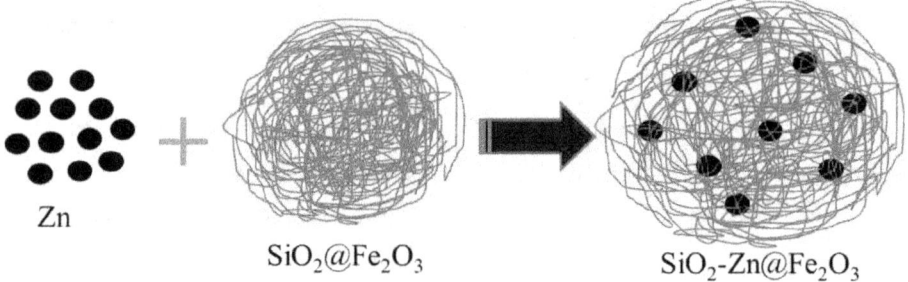

Zn SiO_2@Fe_2O_3 SiO_2-Zn@Fe_2O_3

Scheme 1. Schematic representation of the synthesis of SiO_2-Zn@Fe_2O_3

2.2. Antibacterial activity

Highly drug resistant bacterial species *E. coli* and *S. Aureus* were chosen as test bacteria. The strains were obtained in biotechnology department at USTB.

2.2.1. Antibacterial activity of SiO$_2$-Zn@Fe$_2$O$_3$

To investigate the antibacterial efficacy of SiO$_2$-Zn@Fe$_2$O$_3$, the agar well protocol was applied [33]. The strains of the said bacteria were grown in broth media at 37 C. The Said bacteria individually splashed on petri dishes containing agar media with the help of sterile glass. Then the wells of 6 mm diameter were generated. After that SiO$_2$-Zn@Fe$_2$O$_3$ (1 mg) was suspended in visible light for 80 minutes in water (1 mL) and then carefully poured 50 μl of SiO$_2$-Zn@Fe$_2$O$_3$ into the wells. The plates were subsequently moved and placed in incubator at 37°C for 24 hr. As a negative control, the same technique was applied in darkness.

2.2.2. Minimum Inhibitory Concentration (MIC)

MIC of the synthesized nanomaterials was investigated by modified method [34-36]. According to the procedure 1 mL of specific bacterial solution was taken in different sterile test tubes. Then different concentrations of the said nanofibers (10-80 μg/mL) were added. Following blending, the test tubes were emptied into clean Petri dishes. The Petri dishes were then positioned for 20 h in an incubator at 37 C. A test tube without SiO$_2$-Zn@Fe$_2$O$_3$ was adopted as a negative control.

2.2.3. Reactive Oxygen Species Production

In the presence of 2, 7dichlorodihydrofluorescein diacetate dye, ROS production was examined. For the analysis of ROS in bacterial cells, this essential dye is very precise. Specific amounts of SiO$_2$-Zn@Fe$_2$O$_3$ were hatched along the *E. coli* at 300 rpm for 3 h. After proper incubation, *E. coli* strain hastened *E. coli* cell suspension and washed the precipitated Pellet with saline phosphate buffer (7000rpm8min) (PBS). Subsequently, 1000μL of 15000μM^2, 7di-chloro di-hydro fluoresce in diacetate dye was combined for 1hr with the pellet containing PBS. The dye treated cells were then washed with PBS to extract the said dye from the external surface of the cells. To produce fluorescence images at two wavelengths, i.e. 488 nm lengths of excitation and emission waves, 535 nm, correspondingly, and the fluorescence microscope was used [37].

2.3. DPPH scavenging

The DPPH scavenging of SiO_2-$Zn@Fe_2O_3$ nanofibers was checked according to the modified protocol of Chang et al. [38]. According to this procedure different concentrations of nanostructure (0.2-1.3 mg) was applied on DPPH (1.5 mM). The prepared mixture was put in dark for 20 minutes under constant stirring. The absorbance was fixed at a wavelength of 517 nm by Spectrophotometer agains t methanol. Vit-C was applied as normal. The percent inhibition was calculated by the following formula

(Absorbance control)(Absorbance test)/(Percent inhibition=) (Absorbance control).

3. RESULTS AND DISCUSSION

3.1 High Resonance Transmission Electron Microscopy (HRTEM)

The HRTEM analysis is conducted in order to evaluate the morphology of the specimens and to calculate their particle size. The conclusions of the HRTEM and size distribution investigate of the synthesized SiO_2-$Zn@Fe_2O_3$ composites were shown in figure 1. The surface of the Fe_3O_4 particles has a granular framework signifying that the Fe_3O_4 particles are developed by a combination of nanoparticles with low diameters. The $SiO_2@Fe_3O_4$ composite exhibited structural characteristics comparable to the Fe_3O_4 particles, SiO_2 was coated as a thin chamber mostly around core of Fe_3O_4 and a finer surface of $SiO_2@Fe_3O_4$ was produced particularly in comparison to the surface of Fe_3O_4 owing to the amorphous structure of SiO_2. To prohibit the oxidation of magnetic nanoparticles and the bonding of other nanostructured materials, the amorphous SiO_2 shell observed on the surface of the Fe_3O_4 core is very significant. The SiO_2-$Zn@Fe_2O_3$ composites have a uniform dispersion, such as the $SiO_2@Fe_3O_4$ composite and Zn particles are distributed homogeneously on the $SiO_2@Fe_3O_4$ surface. Particles of Zn are observed on the surface as granular structures. Defects due to the aggregation of the Zn particles have developed in some regions of the SiO_2-$Zn@Fe_2O_3$ composite surface and have produced substantial increases in the diameter of the composite. The examination of particle diameter distribution reported that the composites of SiO_2-$Zn@Fe_2O_3$ have an average diameter of 30 nm and are distributed in the 27 to 35 nm range.

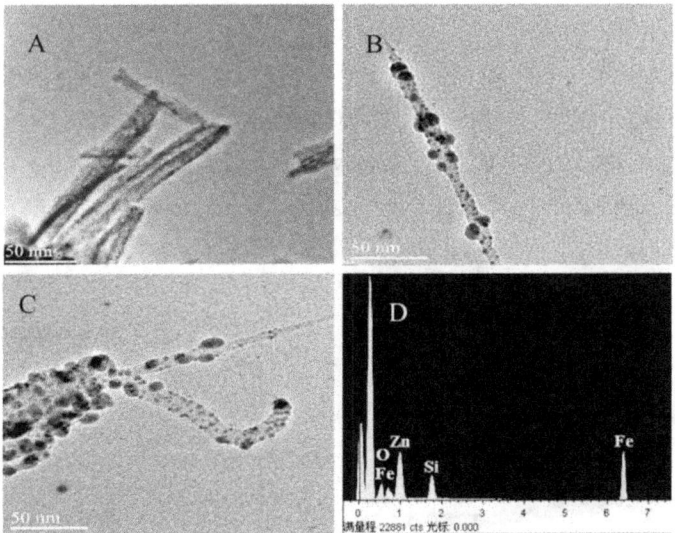

Figure 1. HRTEM analysis of (A) Fe_2O_3 nanofiber (B, C) SiO_2-$Zn@Fe_2O_3$ nanofiber and (D) EDX examination of SiO_2-$Zn@Fe_2O_3$ nanofiber

3.2 Energy dispersive X-ray (EDX) study

In SiO_2-Zn@Fe_2O_3 hybrid nano-fibers, the EDX spectrum confirms the presence of iron, Si, Zn & Oxygen atoms. In comparison, the planning phase of the multiple elements shows the uniform sharing of all elements. It is observed from the findings that SiO_2-Zn@Fe_2O_3has been synthesized successfully.

3.3 X-Ray (XRD) Diffraction Examination

The XRD is used to study the crystalline nature of SiO_2-Zn@Fe_2O_3 nanostructure. Figure 2(A) demonstrates the XRD spectrum of SiO_2 while Figure 2(B) reveals the XRD analysis of SiO_2-Zn@Fe_2O_3. The peak appeared at 22° is for SiO_2. Figure 2 B reveals that all Fe_3O_4 peaks (37°, 42°, 44°, 56°, 72° and 76°) are perfectly indexed to the magnetite form (JCPDS No.19-0629) [39]. Likewise, peaks for Zn are present at (32°, 34°, 36°, 47°, 58°, 64° and 68°) that confirm the successful incorporation of Zn and the result is closely matched with (JCPDS No.34-0394) [40]. Due to the entry of Zn nanoparticles into the SiO_2 shell, the amplitude of the broad peak at 22°is somewhat small which due to the surface area SiO_2 covered by Zn. The percentage mass of Zn in SiO_2-Zn@Fe_2O_3 nano-fibers was taken into account by the ICP process and the value is 3.8 percent wt.

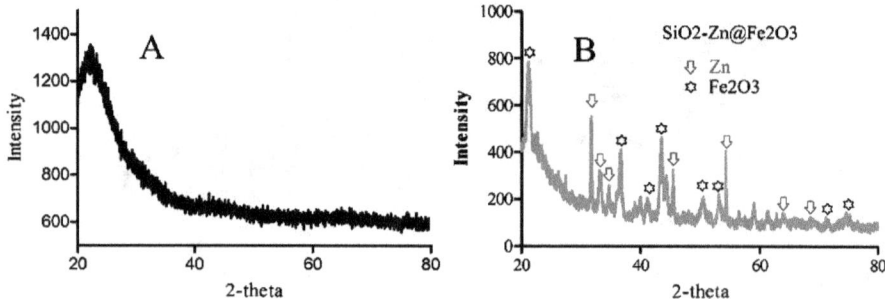

Figure 2. XRD analysis of (A) SiO_2 (B) SiO_2-Zn@Fe_2O_3

3.4. Brunauer–Emmett–Teller (B.E.T)

In addition, BET study (Figure 3A) was applied to investigate the surface area and pore size of the synthesized nanomaterials. The isotherm is type IV that shows that the SiO_2-Zn@Fe_2O_3 nanomaterial is mesoporous and the surface area is 412 m^2/g that is greater than the individual Fe_3O_4 (67 m^2/g). Barrett-Joyner-Halenda method was used to investigate the pore size of the synthesized nanofibers. The pore size of SiO_2-Zn@Fe_2O_3 nanofiber is 2.2 nm which is lower than the individual SiO_2 (3.4 nm).

3.5. Thermo gravimetric analysis (TGA)

TGA of the synthesized nanomaterial was investigated in order to observe the thermal stability at higher temperature. The result is shown in Figure 3B. The outcomes of the experiment showed that the material is highly stable and only less than 20% weight loss was observed at 500 °C temperature.

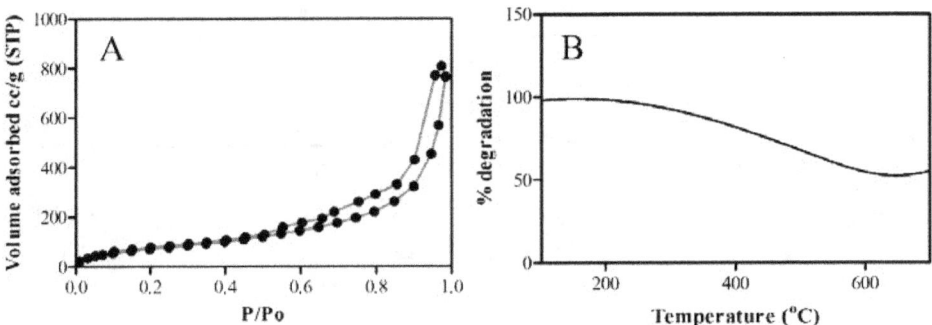

Figure 3. (A) Examination of N_2-adsorption-desorption and (B) Thermo gravimetric analysis of SiO_2-Zn@Fe_2O_3

3.5. X-ray photoelectron spectroscopy (XPS) analysis

XPS techniques have determined the arrangements of multiple surface samples. The XPS spectrum of Si, oxygen, iron, and Zn are seen in Figure 4 (A, B, C, D). The peaks centered at 105.2 eV and 98.1 eV demonstrate the presence of SiO_2 and Si respectively. It is confirmed that Si is found in plus four oxidation state [28]. A peak occurs at 536.7 eV for O, indicating the materials of metal-O [41, 42]. The iron gives 2p well spectrum, which was de-convoluted into 2 peaks based at 718.2 eV, showing Fe^{+3} $2p_{3/2}$ & 729.1eV showing Fe^{+3} $2p_{1/2}$ respectively [43]. A peak also appeared for Zn at 1021.3 eV showing Zn $2p_{3/2}$.

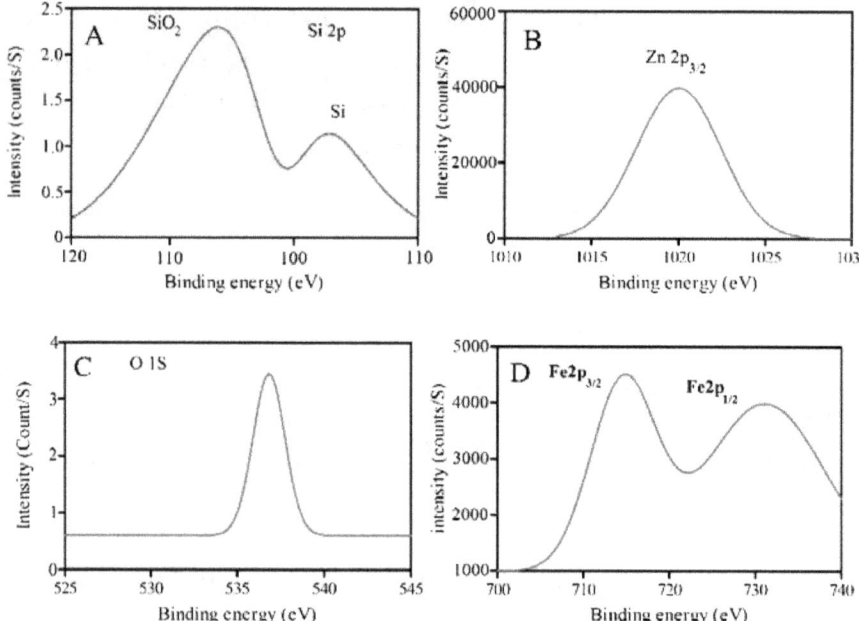

Figure 4. (A, B, C, D) XPS techniques for Si, oxygen, iron, and Zn

13

3.6. Photo degradation of methylene blue

A large amount of the photo catalytic reaction has the tendency and an important role on our environment. Industries discharge organic compounds and dyes in a huge quantity to the environment which have a bad effect on animals, plants and human being. Among all these organic pollutants MB has a large heterocyclic resonating structure, so it is more stable and show more conflict to light & heat, therefore its reduction needs some special environments. MB is a hydrophilic in nature it has 2 different resonating configurations. Owing to its toxic effect of MB on our atmosphere, it is necessary to remove from waste water which is a great challenge for the researchers. For examining the photo catalytic reaction of SiO_2-$Zn@Fe_2O_3$ nanofiber was used in this research. MB illustrates absorption spectra at 614 nm & 663 nm which are used to check the photo catalytic procedure. These spectra of MB at different time interval are shown in (Fig. 4.17A). The results show that the main absorption spectra decreases very quickly in the existence of light &SiO_2-$Zn@Fe_2O_3$, this indicates the reduction of MB. But only in the absence of light this process is very slow. While in the absence of SiO_2-$Zn@Fe_2O_3$,the reaction does not take place. It means that the degradation of MB in the presence of SiO_2-$Zn@Fe_2O_3$ depends on the following three reasons. The first reason is the small and spherical size of Zn and high surface area of nanomaterial. The second is the tendency of light absorption by the synthesized nanofibers. The last cause is its great diffusion deprived of any accumulation. Literature review shows that photo catalytic degradation depends upon the surface area, composition and crystal lattice [44, 45]. According to the proposed mechanism of photocatalytic degradation of MB in the presence of nanofibers, the visible rays strike with the valence electrons of SiO_2-$Zn@Fe_2O_3$ then it gained energy & excited from the valence shell. These emitted electrons are extremely energetic which will be used in the production of OH, where it is liable for degradation of MB.

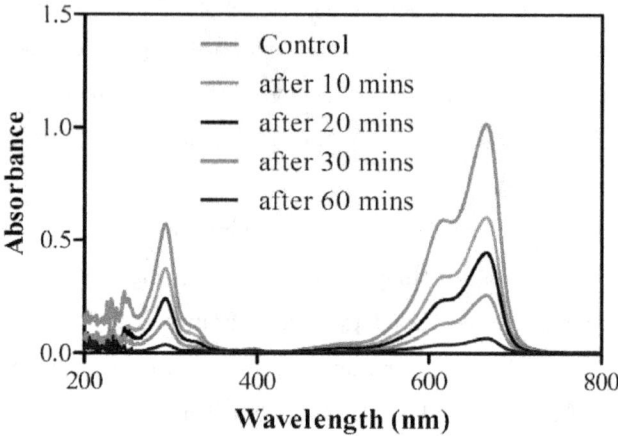

Figure 5. Photocatalytic reduction of MB in the presence of SiO₂-Zn@Fe₂O₃ and
light

3.6.1. Mechanism

Bond dissociation energy plays an important role in the degradation of dyes. The mechanism of dye degradation involves electron transfer from donor to accepter. Dye acts as electron acceptor and nanomaterial that produce ROS which act as electron donor. When light approaches the surface electrons of the nanowire, they absorb this energy from light and jump to high energy level. The energy at excited state of electrons is not enough to produce reactive oxygen species. Thus the excited electrons of ZnNPs will increase the energy of the excited electrons of nanowire and thus they become able to produce reactive oxygen species from water and oxygen molecules.

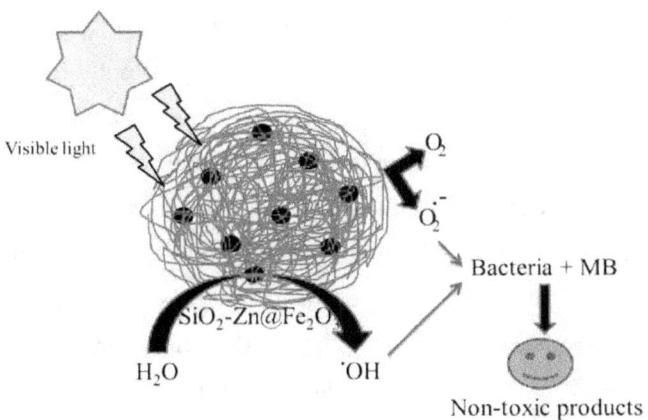

Mechanism 1. Detail mechanism of the photodegradation of Bacteria and Methylene blue in the presence of SiO_2-Zn@Fe_2O_3 nanofibers

3.7. Antibacterial activity of SiO_2-Zn@Fe_2O_3

The antibacterial potential of SiO_2-Zn@Fe_2O_3 nanofibers was tested both in light and dark. The *E. coli* and *S. aureus* were used as test bacteria for the analysis. The bacteria mentioned are the most prevalent bacteria responsible for multiple human infections, such as infection of the urinary tract, bacteremia, cholangitis, etc. [46]. A comparative antibacterial activity analysis was conducted in both dark and visible light, and under these conditions, photoactive SiO_2-Zn@Fe_2O_3 nanofibers exhibited good antibacterial activity. Their antibacterial activity of SiO_2-Zn@Fe_2O_3 nanofibers was significantly improved by visible light irradiation and completely impeded their growth. As seen in Table 1, the antibacterial potency of the synthesized nanomaterial was shown to be much higher in visible light than in dark, as shown in Table 1. Inhibition diameters of irradiated SiO_2-Zn@Fe_2O_3 nanofibers against *E. coli* and *S. aureus* was 13 ($\pm$0.2), 17 ($\pm$0.4) respectively and in dark was 8 ($\pm$0.5) and 12 ($\pm$0.3) respectively. These results have also shown that SiO_2-

Zn@Fe$_2$O$_3$ nanofibers have a pronounced antibacterial activity against these bacteria, both in visible light and in darkness, but most significantly in light. SEM examination was followed of *E. coli* in order to better describe the damage of SiO$_2$-Zn@Fe$_2$O$_3$ nanofibers. The preferred morphology of the bacterium was observed in the presence and absence of SiO$_2$-Zn@Fe$_2$O$_3$ nanofibers, as seen in Fig. 6. Several proposed nanomaterial mechanisms have been tested for antibacterial action. For instance, (1) phosphate and thiol group interaction protein degradation, (2) non-functionalization of bacterial respiratory organs (3) Cell membrane breakdown, (4) decrease in K+ by nanomaterials, resulting in cell transport disturbance, (5) ROS growth [47, 48]. The ROS is extremely reactive and thus capable of destroying the cell membrane, DNA, proteins, etc [49].

Table 1. Antibacterial activity of irradiated and non-irradiated SiO$_2$-Zn@Fe$_2$O$_3$nanofibers

Bacteria	Irradiated SiO$_2$-Zn@Fe$_2$O$_3$	Dark SiO$_2$-Zn@Fe$_2$O$_3$
E. coli	13 (±0.2) mm	8 (±0.5) mm
S. aureus	17 (±0.4) mm	12 (±0.3) mm

3.7.1 Minimum concentration of inhibitory:

The MIC of SiO$_2$-Zn@Fe$_2$O$_3$ nanofibers against the aforementioned bacteria was also investigated. Various amounts (05 to 80µg/mL) of SiO$_2$-Zn@Fe$_2$O$_3$ nanofibers were applied in this study. The MIC for *E. coli* and *S. aureus* was found to be 10 µg/mL and 05 µg/mL respectively as shown in table 2.

17

Table 2. MIC of SiO_2-Zn@Fe_2O_3 nanofibers against *E. coli* and *S. aureus*

Bacteria	SiO_2-Zn@Fe_2O_3(µg/mL)				
	Control 10 05	80	60	40	20
E. coli	+ + +	-	-	-	-
S. aureus	+ - +	-	-	-	-

3.7.2 ROS analysis

The exact mechanism of the antibacterial behavior of nanomaterials is very difficult to analyze. Some suggested pathways have been researched so far. In the present paper, we have attempted to study the exact function of SiO_2-Zn@Fe_2O_3 nanofibers in the presence of visible light against bacteria. They surface electrons gain energy and pass to the excited state when the light approaches in the ground state. Then these high-energy electrons react with O_2 and H_2O_2 and form super oxides. The positive holes formed by the ejection of electrons in the ground state form OH radicals from water molecules. The literature suggests that ZnO has the potential to develop ROS [50]. Such super oxides and OH radicals respond to bacterial cell damage and eventually result in death. In this study, the antimicrobial activity of SiO_2-Zn@Fe_2O_3 nanofibersin microbial cells could be attributable to ROS growth. As a consequence of this, electrons excited from SiO_2-Zn@Fe_2O_3 nanofibers promote the generation of ROS in SiO_2-Zn@Fe_2O_3 nanofibers. The function of these ROSs is to kill the protein and DNA. It is well-proven that in the presence of ROS, the 2, 7-dichlorofluorescindiacetate dye is oxidized into dichlorofluoroscein. The green fluorescence, as seen in Fig.6, is analyzed at 488 nm upon excitation in the presence of SiO_2-$Zn@Fe_2O_3$ nanofibers. In the absence of the said nanomaterials, no fluorescence is detected. This finding specifically shows that ROS is produced in bacterial cells in the presence of SiO_2-$Zn@Fe_2O_3$ nanofibers, and is responsible for their inhibition [51].

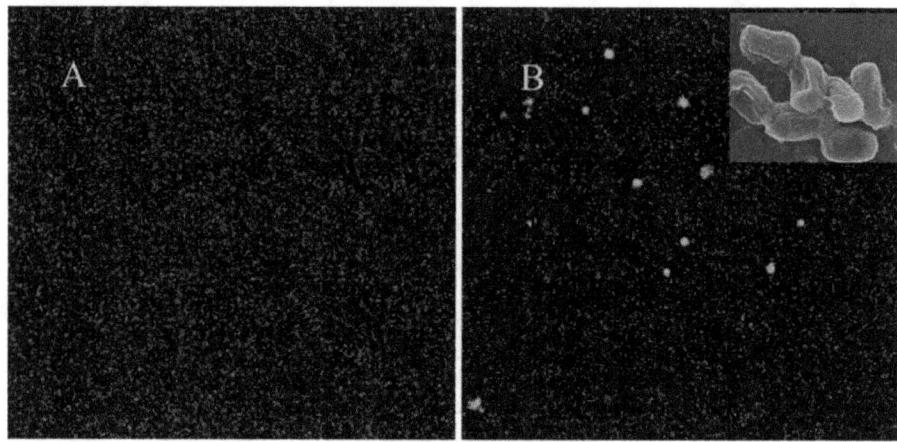

Figure 6. ROS analysis, Zone of inhibition and SEM observation of bacteria in the absence of SiO_2-$Zn@Fe_2O_3$ nanofibers and light (A) and in the presence of SiO_2-$Zn@Fe_2O_3$ nanofibers and light (B)

3.7.3. Irradiation effect on bacterial inhibition

The catalytic efficacy of SiO_2-$Zn@Fe_2O_3$ nanofibers was evaluated by varied irradiation time at room temperature. The study reveals that the inhibition zone of *E. coli* and *S. aureus* was increased, while increasing the duration of irradiation time. After exposure to light irradiation, the electrons present in the outer shell of the synthesized nanocomposite can excite, which leads to enhance the antibacterial activity. The excited electrons were produced super oxide and OH radicals, which destroy the bacterial cells present in the medium. The result shows that, the inhibition percentage is directly proportional to the irradiation time. Time-based inhibition activity indicates that 98% of the tested bacteria were disintegrated after 60 min of irradiation time in the presence of SiO_2-$Zn@Fe_2O_3$ nanofibers. The same conditions were also applied in the presence of individual iron oxide. The results show that the efficiency of SiO_2-$Zn@Fe_2O_3$ nanofibers was several times higher than individual iron oxide. All these results are shown in figure 7.

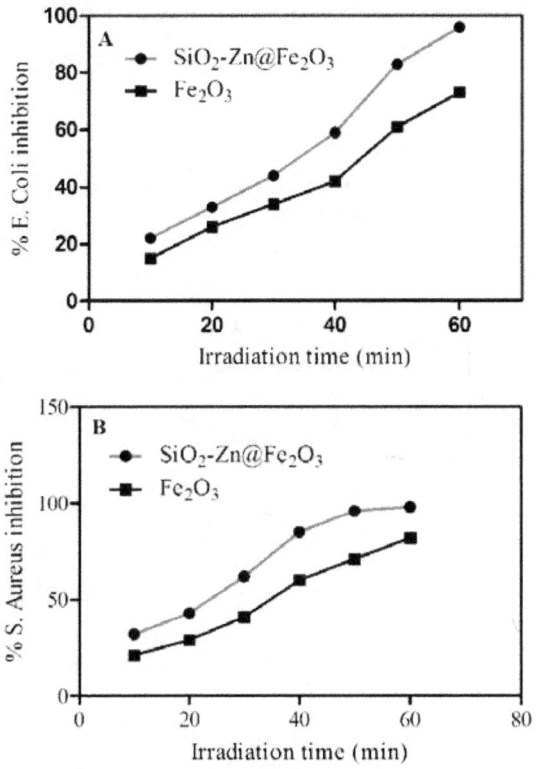

Figure 7. Irradiation effect on the inhibition of bacteria in the presence of SiO_2-$Zn@Fe_2O_3$ nanofibers

3.7.4 Concentration effect of SiO_2-Zn@Fe_2O_3 nanofibers

The photo inhibition potential of SiO_2-Zn@Fe_2O_3 nanofibers against said bacteria was tested at various concentrations (05 to 80 µg/mL) (Figure 8). The experiment was performed at 25°C for 60 minutes of irradiation time. It was observed that the inhibition of each bacteria increases with increased in SiO_2-Zn@Fe_2O_3 nanofibers concentrations. The results showed that both the bacteria inhibited 98% at 60 µg/mL concentration of SiO_2-Zn@Fe_2O_3 nanofibers. The same conditions were also applied in the presence of individual iron oxide. The results show that the inhibition efficacy of SiO_2-Zn@Fe_2O_3 nanofibers was more pronounced than individual iron oxide.

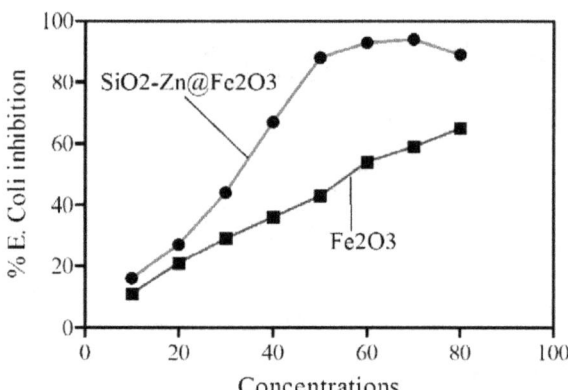

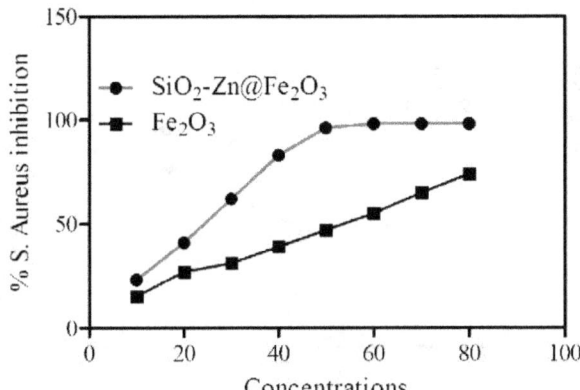

Figure 8. Concentration effect of SiO₂-Zn@Fe₂O₃ nanofibers on the inhibition of bacteria

3.8 Reduction of DPPH

The conversion of DPPH from radical to stable form was tested in the presence of SiO₂-Zn@Fe₂O₃ nanofibers. Vit. C was used as standard. DPPH behavior is demonstrated in Fig. 9. Different concentrations (0.2-1.3 mg) of SiO₂-Zn@Fe₂O₃ nanofibers were fixed for this study. Results showed that with increasing concentrations of nanomaterials, the stability of DPPH radicals also increases. The similar results for selenium have also been reported [52]. The synthesized nanostructures will be a good candidate for reduction of DPPH as its efficacy is closely matched with Vit. C.

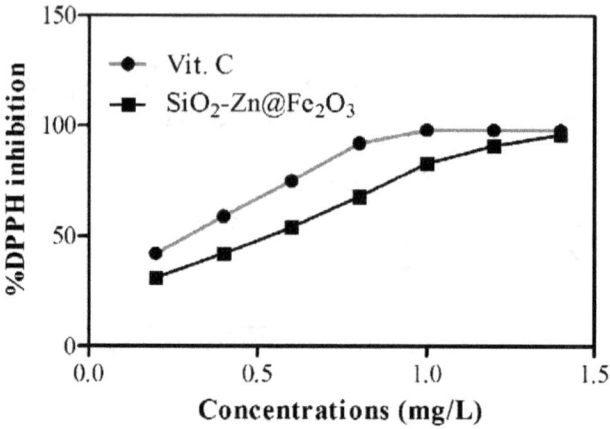

Figure 9. DPPH reduction in the presence of SiO$_2$-Zn@Fe$_2$O$_3$ nanofibers.

Conclusion

A simple hydrothermal process was used to synthesize the mesoporous composite SiO_2-$Zn@Fe_2O_3$. The nanowires obtained have large surface area, desired nanowire length and width. The structure of the synthesized nanomaterial is mesoporous. Two different surfactants were used for the synthesis of the said nanostructures i.e. CTAB and Tween-80. The surface active sites were enhanced by the addition of Zn on the material surface. ZnNPs of a specific size of approximately 20 nm have been observed to disperse very well on the surface of nanostructure. The prepared Nano fibers demonstrate good magnetic behaviors for the dyes and excellent photo catalytic degradation. For anionic dyes, the basic rate constant (k) value is 10 times higher than for cationic dyes, so the decay of anionic dyes is seen to be discernible. In addition, owing to magnetic properties, Fe_2O_3-$Zn@ SiO_2$ can quickly be isolated from the chemical reaction, offering adequate recyclability. A significant volume of Fe_2O_3-$Zn@ SiO_2$ and other helpful nano-materials for environmental & catalytic uses are expected to be prepared from these properties. In both dark and visible light, a comparative antibacterial activity study was performed and, under these conditions, photoactive SiO_2-$Zn@Fe_2O_3$ nanofibers exhibited strong antibacterial activity against *E. coli* and *S. Aureus*. The MIC of SiO_2-$Zn@Fe_2O_3$ nanofibers was also investigated, with 20 µg/mL and 10 µg/mL for *E. coli* and *S. Aureus*, respectively. The ROS is extremely reactive and therefore capable of destroying the cell membrane, DNA, proteins, etc. This finding specifically shows that ROS is produced in bacterial cells in the presence of SiO_2-$Zn@Fe_2O_3$ nanofibers and is responsible for inhibiting them. The catalytic efficacy of SiO_2-$Zn@Fe_2O_3$ nanofibers was evaluated by varying exposure time at room temperature. The study found that the inhibition zone was increased, while increasing the duration of the irradiation time. The findings demonstrate that the efficacy of SiO_2-$Zn@Fe_2O_3$ nanofibers was several times higher than that of individual iron oxides.

Synthesized nanostructures will be a promising match for DPPH reduction, as their efficacy is adversely correlated with Vit. C.

Acknowledment

We are thankful to beijing university of chemical technology for characterization of our samples

References

[1] Ansar, S. M., Haputhanthri, R., Edmonds, B., Liu, D., Yu, L., Sygula, A., & Zhang, D. (2011). Determination of the binding affinity, packing, and conformation of thiolate and thione ligands on gold nanoparticles. *The Journal of Physical Chemistry C, 115*(3), 653-660.

[2] Tsunoyama, H., Ichikuni, N., Sakurai, H., & Tsukuda, T. (2009). Effect of electronic structures of Au clusters stabilized by poly (N-vinyl-2-pyrrolidone) on aerobic oxidation catalysis. *Journal of the American Chemical Society, 131*(20), 7086-7093.

[3] Tsunoyama, H., Sakurai, H., Ichikuni, N., Negishi, Y., & Tsukuda, T. (2004). Colloidal gold nanoparticles as catalyst for carbon− carbon bond formation: application to aerobic homocoupling of phenylboronic acid in water. *Langmuir, 20*(26), 11293-11296.

[4] Wunder, S., Polzer, F., Lu, Y., Mei, Y., & Ballauff, M. (2010). Kinetic analysis of catalytic reduction of 4-nitrophenol by metallic nanoparticles immobilized in spherical polyelectrolyte brushes. *The Journal of Physical Chemistry C, 114*(19), 8814-8820.

[5] Gupta, N., Singh, H. P., & Sharma, R. K. (2011). Metal nanoparticles with high catalytic activity in degradation of methyl orange: an electron relay effect. *Journal of Molecular Catalysis A: Chemical, 335*(1-2), 248-252.

[6] Chakraborty, S., Ansar, S. M., Stroud, J. G., & Kitchens, C. L. (2018). Comparison of colloidal versus supported gold nanoparticle catalysis. *The Journal of Physical Chemistry C, 122*(14), 7749-7758.

[7] Gan, Z., Zhao, A., Zhang, M., Tao, W., Guo, H., Gao, Q., & Liu, E. (2013). Controlled synthesis of Au-loaded Fe 3 O 4@ C composite microspheres with superior SERS detection and catalytic degradation abilities for organic dyes. *Dalton transactions, 42*(24), 8597-8605.

[8] Song, W., Chi, M., Gao, M., Zhao, B., Wang, C., & Lu, X. (2017). Self-assembly directed synthesis of Au nanorices induced by polyaniline and their enhanced peroxidase-like catalytic properties. *Journal of Materials Chemistry C, 5*(30), 7465-7471.

[9] Song, W., Chi, M., Gao, M., Zhao, B., Wang, C., & Lu, X. (2017). Self-assembly directed synthesis of Au nanorices induced by polyaniline and their enhanced peroxidase-like catalytic properties. *Journal of Materials Chemistry C, 5*(30), 7465-7471.

[10] Yao, T., Cui, T., Wang, H., Xu, L., Cui, F., & Wu, J. (2014). A simple way to prepare Au@ polypyrrole/Fe3O4 hollow capsules with high stability and their application in catalytic reduction of methylene blue dye. *Nanoscale, 6*(13), 7666-7674.

[11] Behl, M., & Jain, P. K. (2015). Catalytic Activation of a Solid Oxide in Electronic Contact With Gold Nanoparticles. *Angewandte Chemie, 127*(3), 1006-1011.

[12] Arrii, S., Morfin, F., Renouprez, A. J., & Rousset, J. L. (2004). Oxidation of CO on gold supported catalysts prepared by laser vaporization: direct evidence of support contribution. *Journal of the American Chemical Society, 126*(4), 1199-1205.

[13] Gardner, S. D., Hoflund, G. B., Schryer, D. R., Schryer, J., Upchurch, B. T., & Kielin, E. J. (1991). Catalytic behavior of noble metal/reducible oxide materials

for low-temperature carbon monoxide oxidation. 1. Comparison of catalyst performance. *Langmuir*, *7*(10), 2135-2139.

[14] Oh, S., Kim, Y. K., Jung, C. H., Doh, W. H., & Park, J. Y. (2018). Effect of the metal–support interaction on the activity and selectivity of methanol oxidation over Au supported on mesoporous oxides. *Chemical Communications*, *54*(59), 8174-8177.

[15] Tanaka, K., Padermpole, K., & Hisanaga, T. (2000). Photocatalytic degradation of commercial azo dyes. *Water research*, *34*(1), 327-333.

[16] Fu, F., Dionysiou, D. D., & Liu, H. (2014). The use of zero-valent iron for groundwater remediation and wastewater treatment: a review. *Journal of hazardous materials*, *267*, 194-205.

[17] Ullah, R., & Dutta, J. (2008). Photocatalytic degradation of organic dyes with manganese-doped ZnO nanoparticles. *Journal of Hazardous materials*, *156*(1-3), 194-200.

[18] Gupta, V. K., Jain, R., Mittal, A., Saleh, T. A., Nayak, A., Agarwal, S., & Sikarwar, S. (2012). Photo-catalytic degradation of toxic dye amaranth on TiO2/UV in aqueous suspensions. *Materials Science and Engineering: C*, *32*(1), 12-17.

[19] Devi, L. G., Kumar, S. G., Reddy, K. M., & Munikrishnappa, C. (2009). Photo degradation of Methyl Orange an azo dye by Advanced Fenton Process using zero valent metallic iron: Influence of various reaction parameters and its degradation mechanism. *Journal of hazardous materials*, *164*(2-3), 459-467.

[20] Shahwan, T., Sirriah, S. A., Nairat, M., Boyacı, E., Eroğlu, A. E., Scott, T. B., & Hallam, K. R. (2011). Green synthesis of iron nanoparticles and their application as a Fenton-like catalyst for the degradation of aqueous cationic and anionic dyes. *Chemical Engineering Journal*, *172*(1), 258-266.

[21] Gupta, N., Singh, H. P., & Sharma, R. K. (2010). Single-pot synthesis: plant mediated gold nanoparticles catalyzed reduction of methylene blue in presence of

stannous chloride. *Colloids and Surfaces A: Physicochemical and Engineering Aspects, 367*(1-3), 102-107.

[22] Bao, X., Qin, Z., Zhou, T., & Deng, J. (2018). In-situ generation of gold nanoparticles on MnO2 nanosheets for the enhanced oxidative degradation of basic dye (Methylene Blue). *Journal of Environmental Sciences, 65*, 236-245.

[23] Gong, C., Li, Q., Zhou, H., & Liu, R. (2018). Tiny Au satellites decorated Fe3O4@ 3-aminophenol-formaldehyde core-shell nanoparticles: Easy synthesis and comparison in catalytic reduction for cationic and anionic dyes. *Colloids and Surfaces A: Physicochemical and Engineering Aspects, 540*, 67-72.

[24] Liu, Y., Liu, B., Wang, Q., Liu, Y., Li, C., Hu, W., & Zhang, J. (2014). Three dimensionally ordered macroporous Au/CeO 2 catalysts synthesized via different methods for enhanced CO preferential oxidation in H 2-rich gases. *RSC advances, 4*(12), 5975-5985.

[25] Xiong, J., Song, P., Di, J., & Li, H. (2019). Ultrathin structured photocatalysts: A versatile platform for CO2 reduction. *Applied Catalysis B: Environmental, 256*, 117788.

[26] Veziroglu, S., Kuru, M., Ghori, M. Z., Dokan, F. K., Hinz, A. M., Strunskus, T., & Aktas, O. C. (2017). Ultra-fast degradation of methylene blue by Au/ZnO-CeO2 nano-hybrid catalyst. *Materials Letters, 209*, 486-491.

[27] Khan, Z. U. H., Khan, A., Chen, Y., ullah Khan, A., Shah, N. S., Muhammad, N., & Wan, P. (2017). Photo catalytic applications of gold nanoparticles synthesized by green route and electrochemical degradation of phenolic Azo dyes using AuNPs/GC as modified paste electrode. *Journal of Alloys and Compounds, 725*, 869-876.

[28] Li, G., & Tang, Z. (2014). Noble metal nanoparticle@ metal oxide core/yolk–shell nanostructures as catalysts: recent progress and perspective. *Nanoscale, 6*(8), 3995-4011.

[29] Kaneda, K., & Mizugaki, T. (2019). Design of high-performance heterogeneous catalysts using hydrotalcite for selective organic transformations. *Green Chemistry*, *21*(6), 1361-1389.

[30] Yang, J., Jia, H., Li, B., & Wang, J. (2017, September). Gold/Ceria Nanostructures for Plasmon-Enhanced Catalytic Reactions Under Visible Light. In *ECS Meeting Abstracts* (No. 42, p. 1900). IOP Publishing.

[31] Hadacek, F., & Greger, H. (2000). Testing of antifungal natural products: methodologies, comparability of results and assay choice. *Phytochemical Analysis: An International Journal of Plant Chemical and Biochemical Techniques*, *11*(3), 137-147.

[32] Kahlmeter, G., & Brown, D. F. (2004). Harmonization of antimicrobial breakpoints in Europe—can it be achieved?. *Clinical Microbiology Newsletter*, *26*(24), 187.

[33] Mukharjee Pulok, K. (2002). Quality control of herbal drugs: an approach to evaluation of botanicals. *Business Horizon Publishers, Edition*, *1*, 131-159.

[34] Organization, W. H. O Traditional Medicine Strategy, 2002-2005.

[35] Arakha, M., Pal, S., Samantarrai, D., Panigrahi, T. K., Mallick, B. C., Pramanik, K., & Jha, S. (2015). Antimicrobial activity of iron oxide nanoparticle upon modulation of nanoparticle-bacteria interface. *Scientific reports*, *5*(1), 1-12.

[36] Choi, C. W., Kim, S. C., Hwang, S. S., Choi, B. K., Ahn, H. J., Lee, M. Y., ... & Kim, S. K. (2002). Antioxidant activity and free radical scavenging capacity between Korean medicinal plants and flavonoids by assay-guided comparison. *Plant science*, *163*(6), 1161-1168.

[37] Cheng, F. Y., Su, C. H., Yang, Y. S., Yeh, C. S., Tsai, C. Y., Wu, C. L., & Shieh, D. B. (2005). Characterization of aqueous dispersions of Fe3O4 nanoparticles and their biomedical applications. *Biomaterials*, *26*(7), 729-738.

[38] Jiang, S., Yang, X., Chen, J., Xing, X., Wang, L., & Yu, R. (2016). Microstructure construction and composition modification of CeO 2 macrospheres with superior performance. *Inorganic Chemistry Frontiers*, *3*(1), 92-96.

[39] Liu, Q., Zhang, Z., Liu, B., & Xia, H. (2018). Rare earth oxide doping and synthesis of spinel ZnMn2O4/KIT-1 with double gyroidal mesopores for desulfurization nature of hot coal gas. *Applied Catalysis B: Environmental*, *237*, 855-865.

[40] Chen, G., Song, G., Zhao, W., Gao, D., Wei, Y., & Li, C. (2018). Carbon sphere-assisted solution combustion synthesis of porous/hollow structured CeO2-MnOx catalysts. *Chemical Engineering Journal*, *352*, 64-70.

[41] Hadidi, L., Davari, E., Ivey, D. G., & Veinot, J. G. (2017). Microwave-assisted synthesis and prototype oxygen reduction electrocatalyst application of N-doped carbon-coated Fe3O4 nanorods. *Nanotechnology*, *28*(9), 095707.

[42] Long, Y., Li, J., Wu, L., Wang, Q., Liu, Y., Wang, X., & Zhang, H. (2019). Construction of trace silver modified core@ shell structured Pt-Ni nanoframe@ CeO 2 for semihydrogenation of phenylacetylene. *Nano Research*, *12*(4), 869-875.

[43] Kamat, P. V. (1993). Photochemistry on nonreactive and reactive (semiconductor) surfaces. *Chemical Reviews*, *93*(1), 267-300..

[44] Balazs, N., Mogyorosi, K., Sranko, D. F., Pallagi, A., Alapi, T., Oszko, A., & Sipos, P. (2008). The effect of particle shape on the activity of nanocrystalline TiO2 photocatalysts in phenol decomposition. *Applied Catalysis B: Environmental*, *84*(3-4), 356-362.

[45] Haynes, C. L., McFarland, A. D., Zhao, L., Van Duyne, R. P., Schatz, G. C., Gunnarsson, L., & Käll, M. (2003). Nanoparticle optics: the importance of radia-

tive dipole coupling in two-dimensional nanoparticle arrays. *The Journal of Physical Chemistry B, 107*(30), 7337-7342.

[46] Chamakura, K., Perez-Ballestero, R., Luo, Z., Bashir, S., & Liu, J. (2011). Comparison of bactericidal activities of silver nanoparticles with common chemical disinfectants. *Colloids and Surfaces B: biointerfaces, 84*(1), 88-96.

[47] Morones, J. R., Elechiguerra, J. L., Camacho, A., Holt, K., Kouri, J. B., Ramírez, J. T., & Yacaman, M. J. (2005). The bactericidal effect of silver nanoparticles. *Nanotechnology, 16*(10), 2346.

[48] Morones, J. R., Elechiguerra, J. L., Camacho, A., Holt, K., Kouri, J. B., Ramírez, J. T., & Yacaman, M. J. (2005). The bactericidal effect of silver nanoparticles. *Nanotechnology, 16*(10), 2346.

[49] Ahmad, A., Ullah, S., Ahmad, W., Yuan, Q., Taj, R., Khan, A. U., & Khan, U. A. (2020). Zinc oxide-selenium heterojunction composite: Synthesis, characterization and photo-induced antibacterial activity under visible light irradiation. *Journal of Photochemistry and Photobiology B: Biology, 203*, 111743.

[50] He, W., Jia, H., Cai, J., Han, X., Zheng, Z., Wamer, W. G., & Yin, J. J. (2016). Production of reactive oxygen species and electrons from photoexcited ZnO and ZnS nanoparticles: a comparative study for unraveling their distinct photocatalytic activities. *The Journal of Physical Chemistry C, 120*(6), 3187-3195.

[51] Ahmad, A., Wei, Y., Syed, F., Tahir, K., Rehman, A. U., Khan, A., & Yuan, Q. (2017). The effects of bacteria-nanoparticles interface on the antibacterial activity of green synthesized silver nanoparticles. *Microbial pathogenesis, 102*, 133-142.

[52] Gao, X., Zhang, J., & Zhang, L. (2002). Hollow sphere selenium nanoparticles: their in-vitro anti hydroxyl radical effect. *Advanced Materials*, *14*(4), 290-293

Publisher: Eliva Press SRL

Email: info@elivapress.com

www.ingramcontent.com/pod-product-compliance
Lightning Source LLC
Chambersburg PA
CBHW072046190526
45165CB00018B/1847